WILDLIFE IN BLOOM SERIES

Little Deer

BY AUTHOR & CONSERVATIONIST

LINDA BLACKMOOR

ISBN: 979-8-9904465-7-1 (PRINT)

PUBLISHED BY QUILL PRESS. LINDA BLACKMOOR'S TITLES MAY BE
PURCHASED IN BULK FOR EDUCATIONAL, BUSINESS, FUNDRAISING, OR
SALES PROMOTIONAL USE. FOR INFORMATION, PLEASE EMAIL
HELLO@LINDABLACKMOOR.COM

FIRST PRINT EDITION: 2024

LINDA BLACKMOOR
WWW.LINDABLACKMOOR.COM

SPECIES

Deer encompass over 50 species like red deer, white-tailed deer, sika, elk, fallow, and tiny pudu —the world's smallest deer. They grace forests, grasslands, and mountains across every continent except Antarctica, adapting to diverse habitats from tropical jungles to arctic tundras. Each species exhibits unique traits, such as the moose with its colossal antlers and the reindeer, where both males and females grow antlers.

ANTLERS

Male deer, called bucks or stags, grow antlers made of bone that are shed and regrown annually—a rare trait among mammals. Antlers can grow rapidly, up to an inch per day, covered in a velvety skin supplying blood and nutrients. Used in displays and battles during mating season, they help establish dominance and attract females. This natural cycle symbolizes renewal and the grandeur of the wild.

DEER FACT #3

CAMO

Deer's coats change with the seasons, offering camouflage that helps them blend seamlessly with their environment. In summer, their fur is often reddish-brown, matching vibrant forests, while in winter, it turns gray or dull brown to blend with barren landscapes. Fawns are born with white spots resembling dappled sunlight on the forest floor, providing concealment from predators.

DEER FACT #4

GRACE

Renowned for their agility and speed, deer can run up to 30 miles per hour and leap over obstacles up to 10 feet high with remarkable grace. Their slender legs and powerful muscles enable them to navigate dense forests and open fields silently and swiftly. Specialized hooves provide traction and absorb shock, allowing them to move with silent elegance.

DEER FACT #5

SENSES

With large, sensitive ears and eyes adapted for low-light conditions, deer possess acute senses of hearing and vision. Their eyes are positioned on the sides of their heads, granting a wide field of view to detect predators. Though they perceive colors differently than humans, they are highly sensitive to movement, especially at dawn and dusk. A keen sense of smell further aids in detecting danger.

RUMINANT

As herbivores, deer are ruminants with a specialized four-chambered stomach that allows them to digest tough plant materials. They chew cud —regurgitated, partially digested food—to extract maximum nutrients from leaves, twigs, and grasses. This efficient digestive system enables them to thrive where food quality varies. Their feeding habits play a vital role in maintaining ecosystems.

DEER FACT #7

MIGRATION

Certain deer species, like caribou, undertake epic migrations, traveling up to 3,000 miles annually—the longest of any land mammal. Guided by the changing seasons and the search for food, they traverse tundras, forests, and mountains in vast herds. These journeys are essential for their survival. The endurance and instinct displayed in these migrations are remarkable.

DEER FACT #8

VOCALS

Deer communicate using vocalizations, body language, and scent markings. They may grunt, bleat, or snort to signal alarm, establish dominance, or maintain contact within the herd. Scent glands on their legs, heads, and hooves release pheromones conveying information about identity and reproductive status. This complex communication system helps them navigate social structures.

SOCIAL

While some deer are solitary, many species form social groups called herds for increased protection. Female deer, or does, often gather with their offspring, while males may form bachelor groups or stay solitary except during mating season. Hierarchies are established through subtle interactions and displays of antler size and strength. This social organization enhances their ability to detect predators.

FAWNS

Deer fawns are typically born in spring after a gestation period of six to eight months, depending on the species. They can stand and walk within hours of birth but remain hidden for the first weeks, relying on their spotted camouflage. Mothers nurse and protect them diligently, returning several times a day to feed them. This nurturing period is critical for their growth and survival.

DEER FACT #11

ADAPT

Deer have evolved remarkable adaptations to survive in various habitats. Reindeer, for example, have specialized noses that warm cold air before it reaches their lungs, and their hooves adapt seasonally for better traction on snow or soft ground. Desert-dwelling mule deer conserve water efficiently and are active during cooler parts of the day.

SPOTS

While most deer lose their spots as they mature, species like the sika and fallow deer retain these enchanting patterns into adulthood. The sika deer, native to East Asia, displays white spots on a reddish-brown coat throughout its life, blending with the dappled sunlight of dense forests. Fallow deer, roaming European woodlands, showcase spotted coats and unique palmate antlers.

THREATS

Some deer species face threats from habitat loss, overhunting, and climate change, leading to population declines. Conservation efforts include habitat protection, regulated hunting, and breeding programs for endangered species like the Père David's deer, once extinct in the wild. Public awareness and sustainable practices are essential to ensure deer continue to thrive.